Taking Shape

by Ricardo Ramos
illustrated by Julia Gorton

Copyright © by Houghton Mifflin Harcourt Publishing Company

All rights reserved. No part of this work may be reproduced or transmitted in any form or by any means, electronic or mechanical, including photocopying or recording, or by any information storage and retrieval system, without the prior written permission of the copyright owner unless such copying is expressly permitted by federal copyright law. Requests for permission to make copies of any part of the work should be submitted through our Permissions website at https://customercare.hmhco.com/contactus/Permissions.html or mailed to Houghton Mifflin Harcourt Publishing Company, Attn: Intellectual Property Licensing, 9400 Southpark Center Loop, Orlando, Florida 32819-8647.

Printed in Mexico

ISBN 978-1-328-77217-6

2 3 4 5 6 7 8 9 10 0908 25 24 23 22 21 20 19 18 17

4500675338   A B C D E F G

If you have received these materials as examination copies free of charge, Houghton Mifflin Harcourt Publishing Company retains title to the materials and they may not be resold. Resale of examination copies is strictly prohibited.

Possession of this publication in print format does not entitle users to convert this publication, or any portion of it, into electronic format.

Everywhere I go, I look for shapes.
I like to find shapes in shapes.
Today, I'm going to the city.
What shapes will I find?
I think I will take pictures!

The first thing I see is a billboard.
This shape is a rectangle.
I also see other shapes in it.
Do you see them, too?

Read • Think • Write What shapes do you see within the rectangle? How many do you count?

At our next stop, I see different shapes.
I see a rectangle in the middle.
I see two parts of another shape on the ends.
Do you see them, too?

Read • Think • Write What other shape do the two parts make?

I walk across a sky bridge.
It has many shapes.
Inside those shapes, I see other shapes.
Do you see them, too?

Read • Think • Write What shapes are on the sides of the bridge? What shapes are within those?

I look up at a tall building.
It looks like a cake!
Near the top, I see shapes.
Do you see them, too?

Read • Think • Write What shapes are near the top?

Before going home, I stop at an old building.
It has windows with 8 sides.
In the windows, I see other shapes.
Do you see them, too?

Read • Think • Write What do you call an eight-sided shape?

Responding

Vocabulary

Shapes Inside Shapes

Show

Look at page 3. Draw the billboard and the shapes in the billboard.

Share

Summarize Look at page 3. Tell about the shapes in the billboard.

Write

Look at page 3. Write the names of the shapes in the billboard.